宝宝嘻哈乐学丛书

神奇大自然

罗国庆 陈良萍 编著

山东大学出版社

图书在版编目（CIP）数据

神奇大自然/罗国庆，陈良萍编著．
—济南：山东大学出版社，2014.10
（宝宝嘻哈乐学丛书）
ISBN 978-7-5607-5068-2

Ⅰ．①神…
Ⅱ．①罗… ②陈…
Ⅲ．①自然科学—儿童读物
Ⅳ．①N49

中国版本图书馆CIP数据核字（2014）第149801号

策划编辑：刘森文
责任编辑：刘森文　郑琳琳
封面设计：祝阿工作室

出版发行：山东大学出版社
　　　社　址：山东省济南市山大南路20号
　　　邮　编：250100
　　　电　话：市场部（0531）88364466
经　　销：山东省新华书店经销
印　　刷：济南新先锋彩印有限公司
规　　格：880毫米×1230毫米　1/16
　　　　　4印张　53千字
版　　次：2014年10月第1版
印　　次：2014年10月第1次印刷
定　　价：18.00元

版权所有，盗印必究
凡购本书，如有缺页、倒页、脱页，由本社营销部负责调换

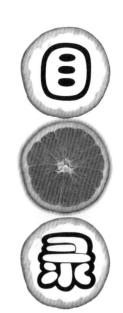

目录

出入迷宫 —— 蜜蜂采蜜 /1
过目不忘 —— 海边沙滩 /2
匹配连线 —— 地形地貌 /5
拼接图片 —— 美妙风景 /6
连接字母 —— 地上的小狗 /7
排列次序 —— 蝴蝶的生命循环 /8
找出影子 —— 展翅飞翔的鸟儿 /9
推断物品 —— 各种球 /10
模仿绘画 —— 鱼儿 /11
出入迷宫 —— 飘落的雪花 /12
与众不同 —— 海洋生物 /13
填充图片 —— 发现隐藏的生物 /14
匹配连线 —— 自然景观 /15
临摹格子 —— 画狮子 /16
亲子手工 —— 拼接恐龙 /17
敏锐观察 —— 找出特别的飞蛾 /19
色彩缤纷 —— 沙漠风光 /20
找出相同 —— 小瓢虫 /21
连接字母 —— 水中的小鸭子 /22
对称涂鸦 —— 画植物 /23
匹配连线 —— 天气相关 /24
拼接图片 —— 海底世界 /25
沿着线走 —— 树上结的果实 /27
找出不同 —— 秋日落叶 /28
出入迷宫 —— 三只小猪找房子 /29

找出错误 —— 冰雪世界 /30
找出错误 —— 树上的东西 /32
认识昆虫 —— 哪些不是昆虫？ /33
对称格子 —— 画稻草人 /34
亲子手工 —— 天气大转盘 /35
推断物品 —— 水果 /37
模仿绘画 —— 画毛毛虫 /38
敏锐观察 —— 长颈鹿身上的东西 /39
认识季节 —— 不同季节的服装 /40
排列次序 —— 小虫子过水坑 /41
出入迷宫 —— 蚂蚁进洞 /42
色彩缤纷 —— 飞机与帆船 /43
找出影子 —— 城堡 /44
出入迷宫 —— 大象捡西瓜 /45
与众不同 —— 优雅的紫花儿 /46
自制小书 —— 植物迷你书 /47
敏锐观察 —— 每种狗有多少只 /49
色彩缤纷 —— 各种颜色的蝴蝶 /50
认识季节 —— 不同季节的事物 /51
对称涂鸦 —— 画食物 /52
填充图片 —— 发现隐藏的鲨鱼 /53
动物分类 —— 哺乳类、鸟类、昆虫类和鱼类 /54
找出不同 —— 春日美景 /55
认识昆虫 —— 哪些是有益的昆虫？ /56

答案与提示 /57

出入迷宫 — 蜜蜂采蜜

这只蜜蜂喜欢红色、粉色的花朵。这只蜜蜂必须从起点飞到终点,并且要采每一朵红色或粉色的花朵,每朵花儿只能采一次。蜜蜂可以向上、向下、向左和向右飞,但不能斜对角飞,也不能飞到白色、绿色、蓝色、黄色的花朵上。请你帮这只忙碌的小蜜蜂指出正确的采蜜路线。(The bee likes red and pink flowers. He must go from the start to the end, and visit each red or pink flower only once. He can fly up, down, left and right, but not diagonally. The bee can not fly over any white, green, blue, yellow flowers. Help the busy bee to collect nectar from all the red and pink flowers.)

过目不忘 — 海边沙滩

仔细观看下面的图片一分钟,然后翻到下一页回答问题。(Look carefully at the picture for one minute, then turn the page to answer some questions.)

前两页的图片内容你能记得多少？不许偷看哦！
（How many things can you remember from the picture on the previous two pages? No peeking!）

1. 天上有几只鸟儿在飞翔？
（How many birds are flying in the sky?）

2. 天上有几只热气球？
（How many hot-air balloons are in the sky?）

3. 沙滩上有什么动物？
（What animal is on the beach?）

4. 水中有什么动物？
（What animal is in the water?）

5. 滑梯是什么颜色？
（What color is the slide?）

6. 船是在水中吗？
（Is the boat in the water?）

7. 树上有多少个椰子？
（How many coconuts are there on the trees?）

8. 球是什么颜色的？
（What color is the ball?）

9. 沙滩上有遮阳伞吗？
（Are there any beach umbrellas on the beach?）

匹配连线 —— 地形地貌

请你将下面的图片与对应的词语用线连起来。（Draw a line from each picture to the matching word.）

丘陵
(hills)

田野
(field)

高山
(mountains)

沙漠
(desert)

瀑布
(waterfall)

小路
(path)

池塘
(pond)

湖泊
(lake)

海洋
(sea)

洞穴
(cave)

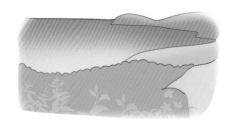
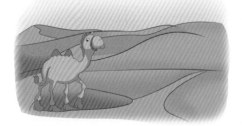

拼接图片 — 美妙风景

请你将下面的图片拼接成一幅完整的图像，把正确的序号填在方框中。（Fill the numbers in the blank to make a perfect picture.）

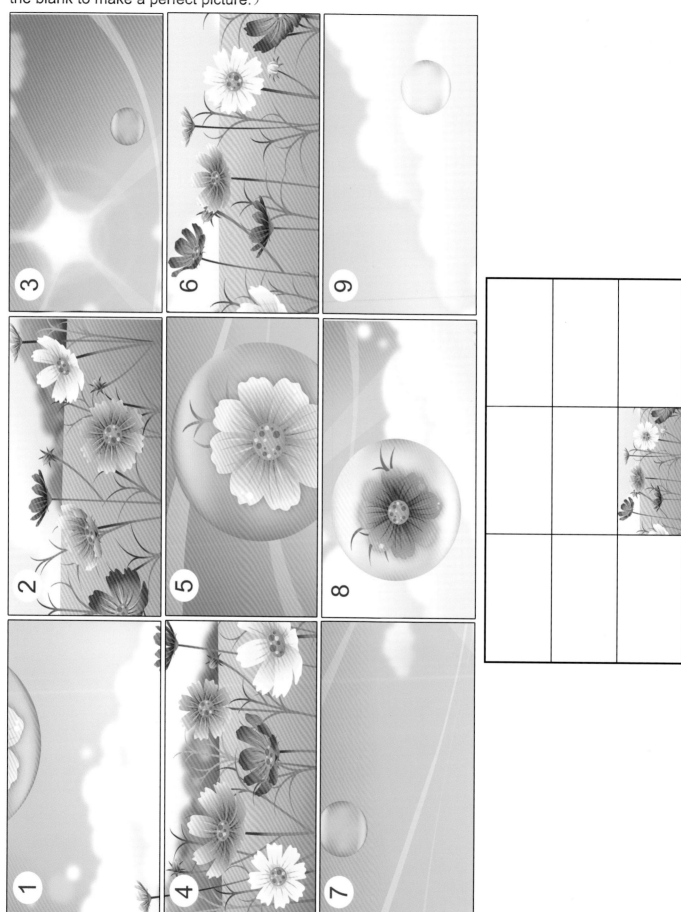

连接字母——地上的小狗

请你按照从"a"到"z"的顺序连接每个字母，同时大声说出每个字母。（While saying each letter aloud, draw a line from "a" to "z" to connect the letters in alphabetical order.）

排列次序 —— 蝴蝶的生命循环

下面是关于蝴蝶生命循环的图片，包括卵、毛毛虫、蛹和成虫四个阶段。请将正确的数字填写在下面的圆圈中。（The following pictures describe the life cycle of a butterfly, including the egg, caterpillar, pupa, and adult. Fill the correct numbers in the circles below.）

这是成虫——蝴蝶。
(This is an adult — butterfly.)

1

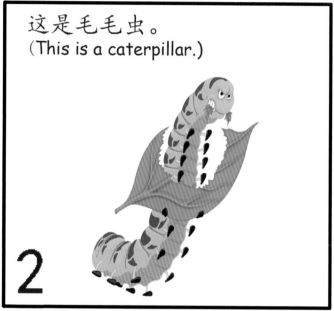

这是毛毛虫。
(This is a caterpillar.)

2

这是蛹。(This is a pupa.)

3

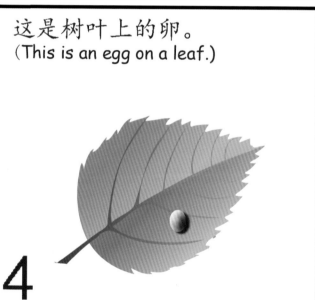

这是树叶上的卵。
(This is an egg on a leaf.)

4

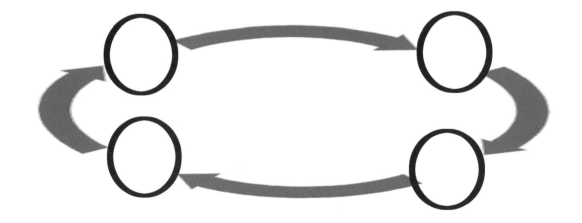

找出影子 — 展翅飞翔的鸟儿

选出与方框中的图片能够精确匹配的影子。(Choose the shadow that matches the picture in the square exactly.)

推断物品 — 各种球

阅读下面的小故事，然后从图片中圈出正确的物体。（Please read the short story below and then circle the correct object in the pictures.）

一个男孩想与朋友们在外面玩游戏。他的朋友让他带一个球出去玩，并且对他说想玩用脚踢的游戏，可以将球踢进球网。这个男孩应该带哪种球呢？（A boy wants to play a game outside with some friends. His friends ask him to bring a ball out to play. They tell him they want to play a game where they can kick a ball around with their feet and then try and kick the ball into a net. Which ball should the boy bring?）

模仿绘画 — 鱼儿

你能遵循下面的步骤画出鱼儿吗？拿起蜡笔，试试吧！（Can you follow the steps below to draw a fish? Try it with your crayon!）

第一步：画一个三角形作为鱼的身体。
Step 1: Draw a triangle for the fish's body.

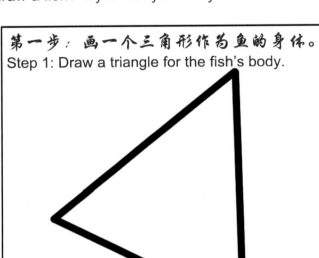

第二步：把三角形的身体变成弧线。
Step 2: Make the curved lines.

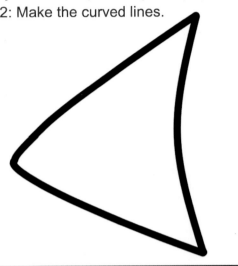

第三步：画尾巴。
Step 3: Draw a tail.

第四步：画眼睛和嘴巴。
Step 4: Draw an eye and a mouth.

第五步：画一条弧线形成鱼的头。
Step 5: Make a curved line for the fish's head.

第六步：涂色。
Step 6: Color the picture.

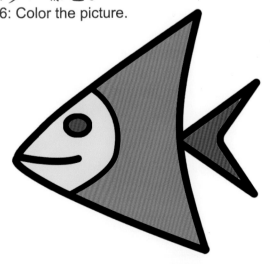

出入迷宫 — 飘落的雪花

请你画出从起点到终点的路径。（Draw a path from the start to the end.）

与众不同 — 海洋生物

下面三幅图中有一幅图与其他图不同，请将这幅不同的图找出来。（Find the picture which is different from the others.）

填充图片 — 发现隐藏的生物

用棕色填充下图中的所有三角形，就可以发现这是什么动物了。（Color brown all the triangles of the following figure to reveal what this animal is.）

匹配连线 — 自然景观

请你将每一幅图片与相匹配的词语连线。（Draw a line from each picture to the matching word.）

天空
(sky)

火
(fire)

森林
(forest)

灌木丛
(bush)

花朵
(flowers)

月亮与星星
(moon and stars)

树
(tree)

雨滴
(raindrop)

彩虹
(rainbow)

树叶
(leaf)

太阳
(sun)

临摹格子 — 画狮子

将下边网格图片临摹到上边的网格中。(Copy the bottom panels into the top panels, square by square.)

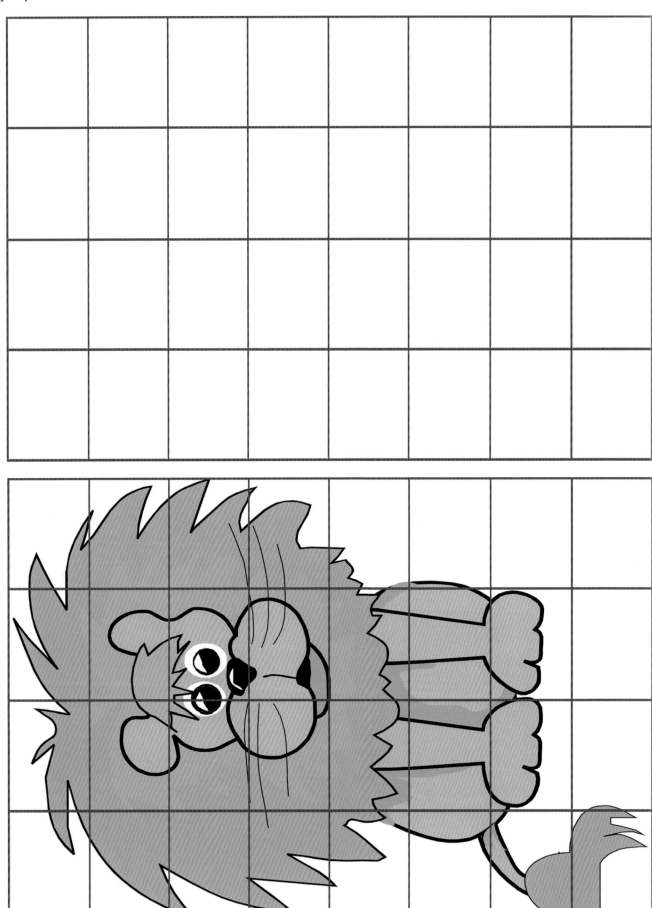

亲子手工——拼接恐龙

请你将本页剪下来,并粘贴在卡片纸上。然后用剪刀沿着图形边线剪开卡片纸,最后将各部分拼接成一只恐龙。(Cut out this page. Glue it onto a piece of cardboard. Cut the cardboard along the edges of each piece. Put them together to make a dinosaur.)

亲子手工

拼接恐龙

(Let's make a dinosaur)

敏锐观察 — 找出特别的飞蛾

请家长根据下面的提示，帮孩子从下图中找出这只特别的飞蛾。(Help your child find the special moth using the clues below.)
- 它在有其他一只飞蛾的行中。(It is in a row that has another moth.)
- 它在有两只甲虫的列中。(It is in a column that has two beetles.)
- 它的右侧没有飞蛾。(There is no moth directly to the right of it.)

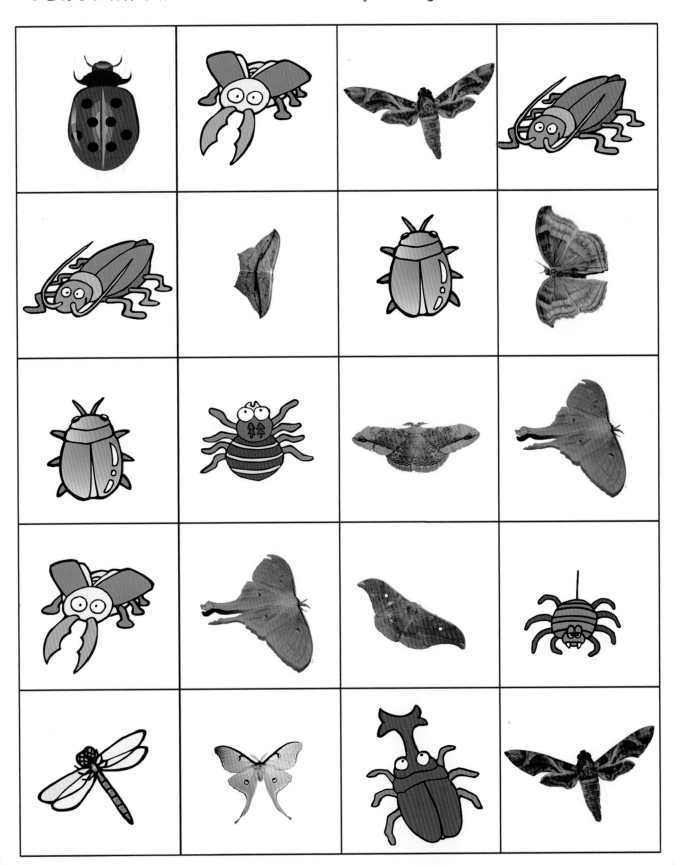

色彩缤纷 — 沙漠风光

请你随意给沙漠里的景物涂色。(Color the objects in desert as you wish.)

找出相同 — 小瓢虫

与方框中的瓢虫完全相同的是哪一只？（Which ladybug is exactly the same as the one in the square box？）

连接字母 — 水中的小鸭子

请你按照从"a"到"w"的顺序连接字母，同时大声说出每个字母。（While saying each letter aloud, draw a line from "a" to "w" to connect the letters in alphabetical order.）

对称涂鸦 — 画植物

利用对称性，来完成树叶和花朵。（Use symmetry to complete the leaf and flower.）

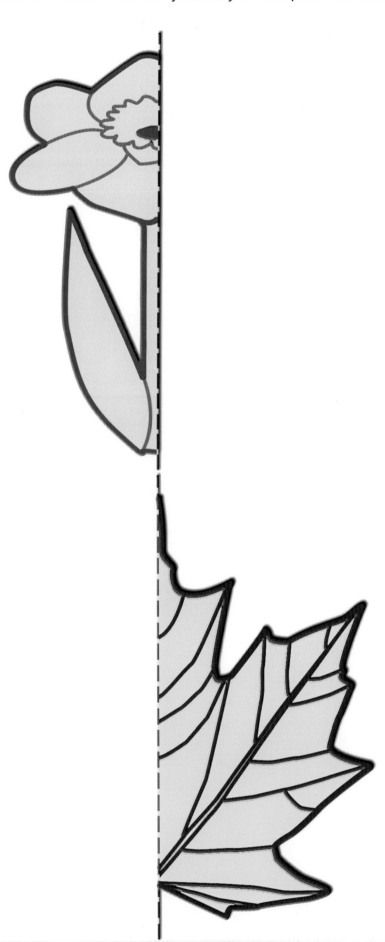

匹配连线 — 天气相关

请你将下面每一幅图片与相匹配的词语连线。（Draw a line from each picture to the matching word.）

刮风的
(windy)

多云的
(cloudy)

下雪的
(snowy)

下雨的
(rainy)

闪电
(lightning)

雪花
(snowflake)

温度计
(thermometer)

晴朗的
(sunny)

拼接图片 — 海底世界

请你将下面的图片剪下来，按照相对应的字母放入下面格子的空白处。（Cut out the pieces below and put them into the proper place in the grid.）

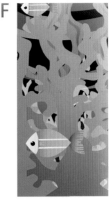
F

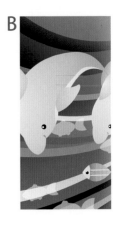
B

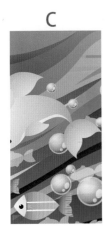
C

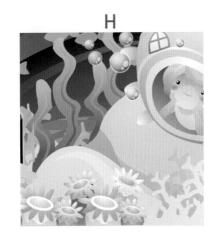
H

I

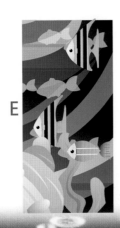
E

G

A

D

25

拼接图片

海底世界

(Let's make an underwater world)

沿着线走 — 树上结的果实

请你沿着细线看看每种树结什么果实。 (Follow the trails to see which tree the fruit grew on.)

找出不同 — 秋日落叶

请你找出下面两幅图片的八个不同之处。（Find the eight differences between the following two pictures.）

出入迷宫 — 三只小猪找房子

请你帮助三只小猪找到去房子的路线。(Help the three little pigs find the way to the house.)

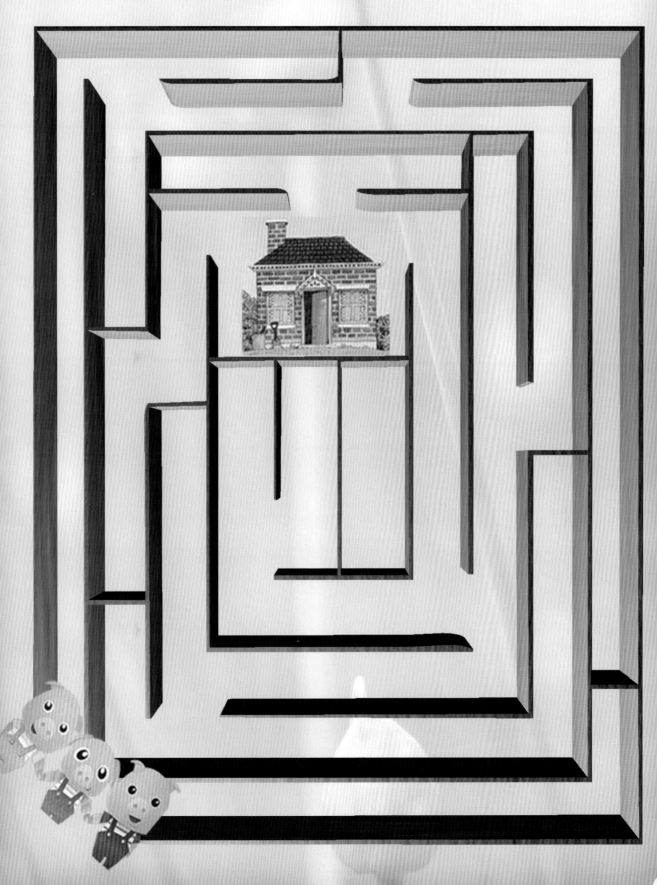

找出错误 — 冰雪世界

在好玩的冬季场景中，找出并圈出八处错误的地方。(Find and circle the eight things that don't match this winter fun scene.)

31

找出错误 — 树上的东西

勾选那些长在树上的东西。(Tick those things that grow on trees.)
叉掉那些长在地下的东西。(Cross out the things that grow on plants under ground.)
圈选那些不是长在树上也不是长在地下的东西。(Circle those things that do not grow on trees and plants under ground.)

认识昆虫——哪些不是昆虫?

请把下面的图片和相对应的词语用线连起来。再圈选出不是昆虫的小动物。(Draw a line from each picture to the matching word. Then circle the bugs that are not insects.)

温馨提示:虿读 mǐn。昆虫都有六条腿。蝎子和蜘蛛有八条腿,蜈蚣号称百足虫,蚯蚓没有腿,故蝎子、蜘蛛、蜈蚣、蚯蚓都不是昆虫。

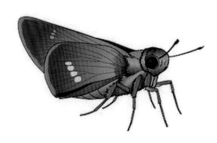

水虿
(water skipper)

飞蛾
(moth)

毛毛虫
(caterpillar)

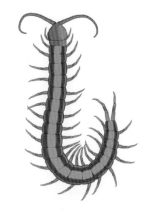

蚂蚁
(ant)

蝉
(cicada)

蚯蚓
(earthworm)

蝎子
(scorpion)

蜘蛛
(spider)

蜈蚣
(centipede)

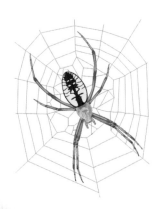

蚊子
(mosquito)

对称格子 — 画稻草人

请你利用格子将稻草人画完整。（Finish off the picture of scarecrow using the grid to help you.）

亲子手工 — 天气大转盘

请你沿着圆圈外缘把转盘的底和转盘的顶剪下来，并剪掉转盘的顶中灰色的区域。再用一个曲别针穿过转盘的顶和转盘的底的中心，将底和顶都连在一起。(Cut off the "wheel bottom" and the "wheel top" along the outer circle and cut out the gray area in the "wheel top". Attach the "wheel top" to the "wheel bottom" with a paper clip through the centers of the two circles.)

当你旋转转盘的时候，就会依次显示八种天气相关（暴风雨、雨滴、刮风、多云、晴朗、雪花、下雪和下雨）图像和名称。(When you spin the wheel, the picture and word are displayed one at a time: storm, raindrop, windy, cloudy, sunny, snowflake, snowy, and rainy.)

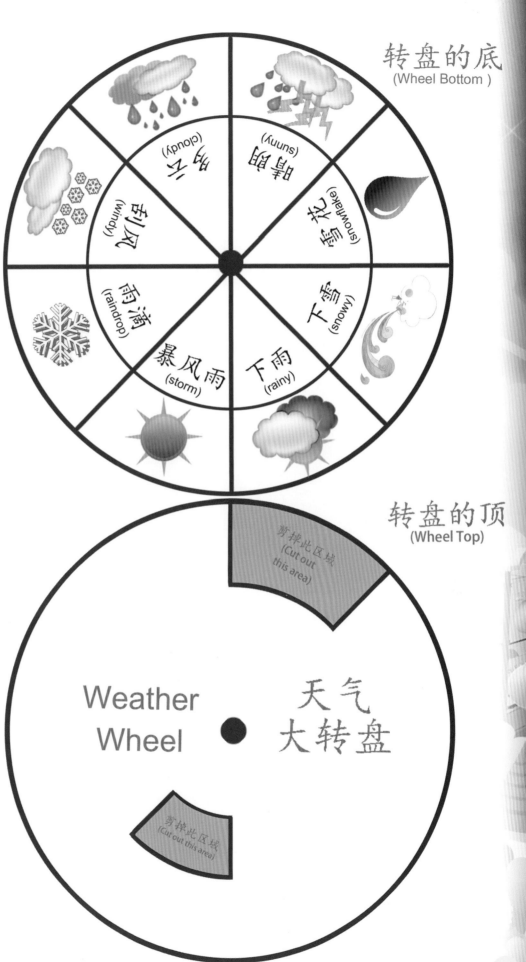

亲子手工

天气大转盘

(Let's make a weather wheel)

推断物品 — 水果

在每个不符合线索的物体上画叉，然后圈选出正确的物体。（Draw a "X" on each object that doesn't fit the clues. Then circle the correct object.）

读下面的线索，推断出哪个水果是正确的答案。（Read the clues below to figure out which fruit is the correct one.）

线索 1：这个水果不是红色的。
（Clue 1: The fruit is not red.）
线索 2：这个水果不是最小的。
（Clue 2: The fruit is not the smallest.）
线索 3：这个水果不是圆形的。
（Clue 3: The fruit is not round.）
线索 4：这个水果不是酸的。
（Clue 4: The fruit is not sour.）
线索 5：这个水果不是棕色的。
（Clue 5: The fruit is not brown.）

模仿绘画 — 画毛毛虫

你能遵循下面的步骤画出毛毛虫吗？拿起蜡笔，试试吧！（Can you follow the steps below to draw a caterpillar? Try it with your crayon!）

第一步：写字母C。 Step 1: Write the letter C. C	第二步：写更多的字母C。 Step 2: Write more letter C's. 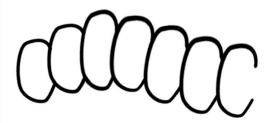
第三步：加上几个O，作为脑袋、眼睛、嘴巴。 Step 3: Add some O's for the head, eyes, and mouth. 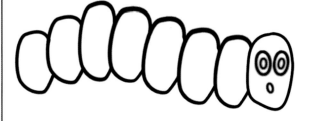	第四步：增加腿。 Step 4: Add legs. 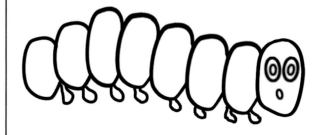
第五步：增加触角。 Step 5: Add antennas. 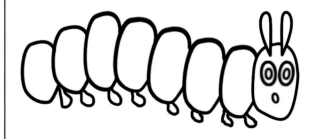	第六步：涂色。 Step 6: Color the picture. 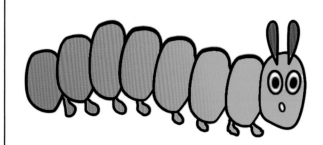

敏锐观察 — 长颈鹿身上的东西

你能在下面的长颈鹿身上找出隐藏的十二个物体吗？分别是：房子、蝴蝶、星星、袜子、书、心形、问号、大写字母H、小写字母c、灯、手提包和茶壶。(Can you find the twelve items hiding in the giraffes? Look for: house, butterfly, star, sock, book, heart, question mark, capital h, letter c, lamp, handbag and teapot.)

认识季节 — 不同季节的服装

属于春季 A 的服装有：_____
(Find the items you wear in Spring A: _____)

属于夏季 B 的服装有：_____
(Find the items you wear in Summer B: _____)

属于秋季 C 的服装有：_____
(Find the items you wear in Autumn C: _____)

属于冬季 D 的服装有：_____
(Find the items you wear in Winter D: _____)

注：春季和秋季的服装可能相同，故答案不唯一。

排列次序 —— 小虫子过水坑

下面是一则小故事。读这个故事并给故事中的事件排序，然后在下面圆圈中填入正确的数字。（The following is a little story. Read it, then fill the correct number in the circles below to sequence the events in order.）

1. "我要乘船。"它就这么做了。
("I will go on a boat." And it did.)

2. 它来到一个很深很宽无法穿过的水坑边。
(It came to a puddle that was too deep and too wide to cross.)

3. 小虫子四下张望。"我知道怎么办了。"它说。
(The little bug looked around. "I know what to do." It said.)

4. 小虫子在去看望朋友的路上。
(The little bug was on its way to visit a friend.)

当小虫子回来的时候会怎么办？（What will the little bug do when it comes back?）

出入迷宫 — 蚂蚁进洞

这只小蚂蚁忘记了回家的路。请你帮这只小蚂蚁找到它的家吧。(This little ant gets lost. Help him find his home.)

找出影子 — 城堡

请你选出与方框中的图片能够精确匹配的影子。(Choose the shadow that matches the pitcture in the square exactly.)

出入迷宫——大象捡西瓜

大象的朋友们在吃西瓜呢。请你帮大象快点回去与朋友们一起吃西瓜吧,别忘了顺便捡起路上的西瓜带回去。(The elephant's friends are eating watermelons. Find the way for the elephant from the start to the end, collecting the watermelons on the road.)

与众不同 —— 优雅的紫花儿

下面六幅图中有一幅图与其他图不同，请将这幅不同的图找出来。（Find the picture which is different from the others.）

自制小书 — 植物迷你书

按照背面介绍的方法，制作一本关于植物的迷你书，并读一读。（Make a mini-book about plants according to the instructions on the next page. Then read it.）

温馨提示：请让爸爸妈妈帮你一起完成。

制作迷你书的方法
(How to Make Your Mini-Book)

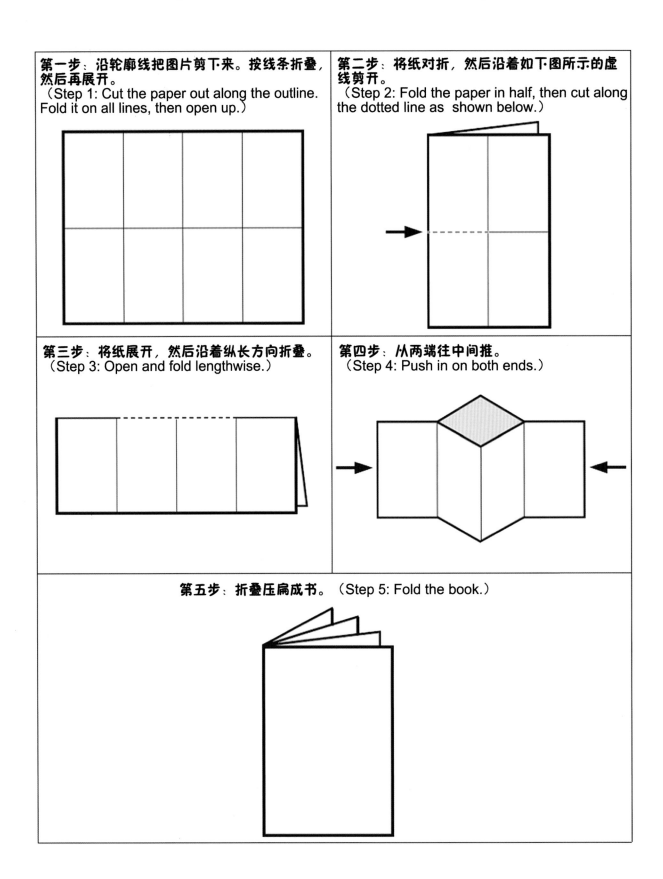

敏锐观察 — 每种狗有多少只

1. 有多少不同种类的狗？（How many different kinds of dogs do you see?）
2. 有斑点的狗多，还是无斑点的狗多？（Are there more spotted or plain dogs?）
3. 有一只狗与其他所有狗都不同，是哪一只？（There is only one dog which is different from all of the others. Which one is it?）

49

色彩缤纷 —— 各种颜色的蝴蝶

1. 将中央方框中的蝴蝶补画完整。并将其涂成绿色。（Finish drawing the butterfly in the center square. Color it green.）
2. 将顶部中间方框中的蝴蝶涂成蓝色，其中的斑点涂成灰色。（Color the butterfly in the top middle square blue with gray spots.）
3. 将中部左侧方框中的蝴蝶涂成棕色，其中的斑点涂成红色。（Color the butterfly in the center left square brown with red spots.
4. 将中部右侧方框中的蝴蝶涂成紫色。（Color the butterfly in the center right square purple.）
5. 将底部中间方框中的蝴蝶涂成红色，其中的斑点涂成白色。（Color the butterfly in the center bottom square red with white spots.）
6. 将顶部右侧方框中的蝴蝶涂成粉红色。（Color the butterfly in the top right square pink.）
7. 将顶部左侧方框中的蝴蝶涂成黄色。（Color the butterfly in the top left square yellow.）
8. 将底部右侧方框中的蝴蝶涂成灰色。（Color the butterfly in the bottom right square gray.）
9. 将底部左侧方框中的蝴蝶涂成橘红色。（Color the butterfly in the bottom left square orange.）
10. 在灰色蝴蝶上画一个红色圆圈。（Draw a red circle around the gray butterfly.）

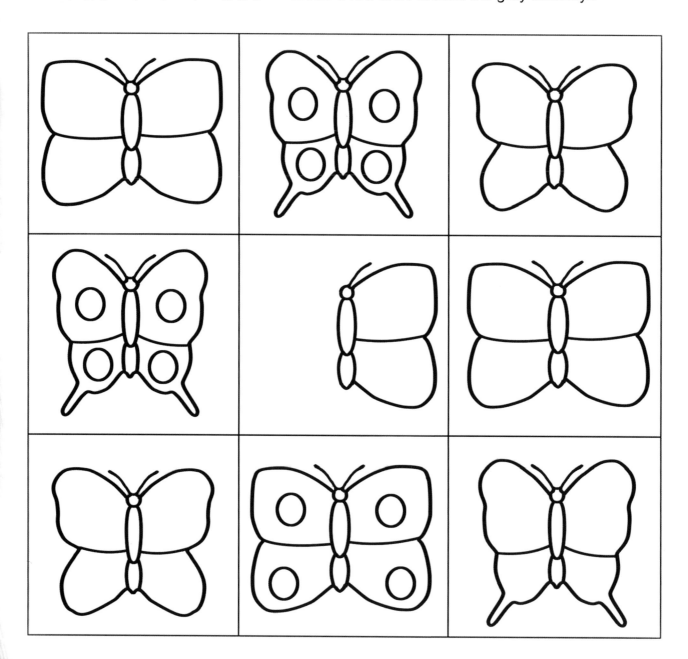

认识季节 — 不同季节的事物

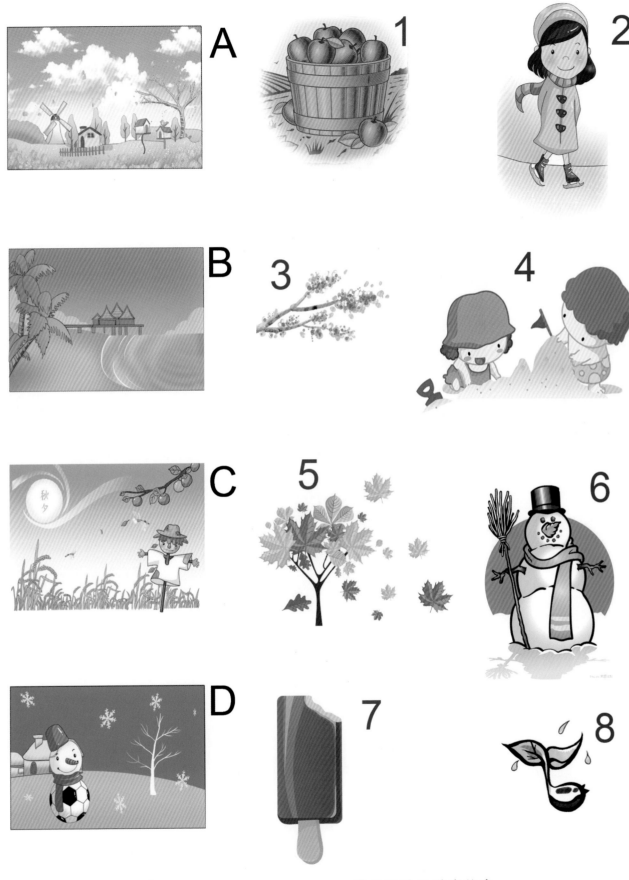

属于春季 A 的事物有：_____
(Find the objects from Spring A: _____)

属于夏季 B 的事物有：_____
(Find the objects from Summer B: _____)

属于秋季 C 的事物有：_____
(Find the objects from Autumn C: _____)

属于冬季 D 的事物有：_____
(Find the objects from Winter D: _____)

对称涂鸦 — 画食物

利用对称性,把下面的食物画完整。(Use symmetry to finish the drawing below.)

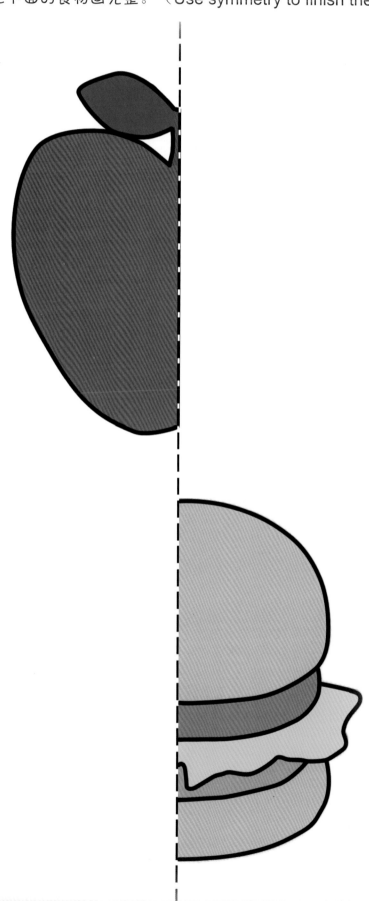

填充图片 — 发现隐藏的鲨鱼

用铅笔将下面列出的方块涂色。完成之后你就能看到一只鲨鱼了。(Use a pencil to shade the squares listed below. When you finish, you will have a picture of a shark.)

- A行: 2、3、4、11、12列。
- B行: 4、5、6、9、10、11列。
- C行: 5、6、7、8、9列。
- D行: 6、7、8列。
- E行: 5、6、7、8、9、10列。
- F行: 5、6、7、8、9、10、11列。
- G行: 4、5、6、7、8、9、10、11、12列。
- H行: 4、5、6、7、8、9、10、11、12、13、14列。
- I行: 4、5、6、7、8、9、10、11、12、13、14、15、16列。
- J行: 4、5、6、7、8、9、10、11、12、13、14、15、16、17、18列。
- K行: 3、4、5、6、7、8、9、10、11、12、13、14、15列。
- L行: 1、2、3、4、5、6、7、8、9、10、11、12、13、14、15列。
- M行: 5、6、7、8、9、10、11、12、13、14、15列。
- N行: 6、7、8、9、10、13、14列。
- O行: 6、7、9、10、11、12、13、14列。
- P行: 9、10、11、12列。
- Q行: 10、11列。

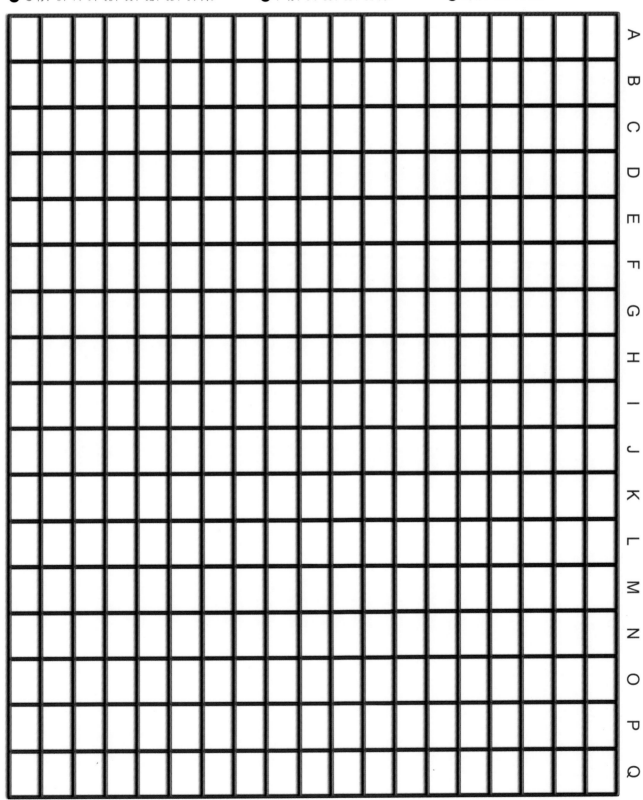

动物分类 — 哺乳类、鸟类、昆虫类和鱼类

将动物按照哺乳类、鸟类、昆虫类、鱼类进行分类。（Classify the animals as mammals, birds, insects, or fish.）

温馨提示：请让爸爸妈妈帮你一起完成。

哺乳类
(mammals)

鸟类
(birds)

昆虫类
(insects)

鱼类
(fish)

找出不同 — 春日美景

找出这两幅图片的六个不同之处。（Find the six differences between these two pictures.）

认识昆虫 —— 哪些是有益的昆虫？

请你把下面的图片与相对应的词语用线连起来。再圈选出有益的昆虫。（Draw a line from each picture to the matching word. Then circle the beneficial insects.）

蜻蜓
(dragonfly)

蟋蟀
(cricket)

蚂蚱
(grasshopper)

蚊子
(mosquito)

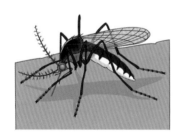

瓢虫
(ladybug)

蝴蝶
(butterfly)

蜜蜂
(bee)

螳螂
(mantis)

苍蝇
(fly)

天牛
(longicorn)

答案与提示

P1

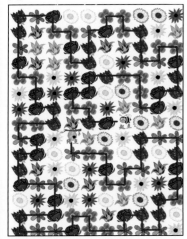

P4

1. 3只（Three）
2. 2只（Two）
3. 一只小狗（A dog）
4. 一只海豚（A dolphin）
5. 粉红色（Pink）
6. 不是（No）
7. 4个（Four）
8. 蓝色、黄色和红色（Blue, yellow and red）
9. 是的（Yes）

P5

P6

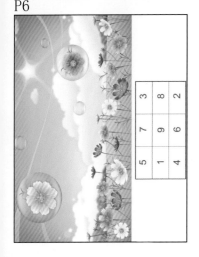

P7

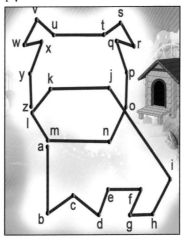

P8

P9

P10

P12

57

P13
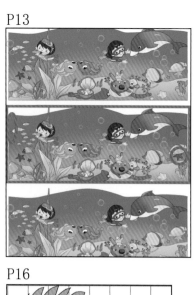

P14

P15
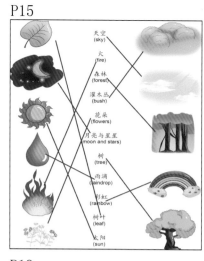

P16

P17

P19
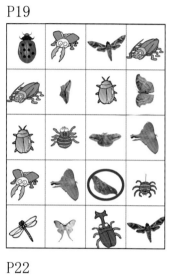

P20

P21

P22

P23

P24
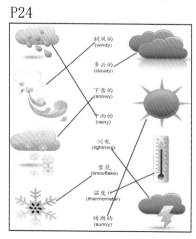

P25

P27

P28

P29

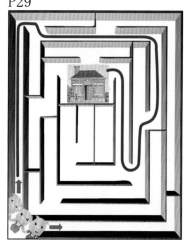

P30

P32

P33

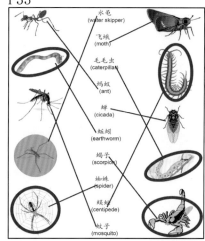

P34

P37

P39

P40

因春季和秋季的服装可能相同，故答案不唯一。
属于春季 A 的服装有：8、11、15。
属于夏季 B 的服装有：2、4、6、7、12。
属于秋季 C 的服装有：13、14。
属于冬季 D 的服装有：1、3、5、9、10。

P41

P42

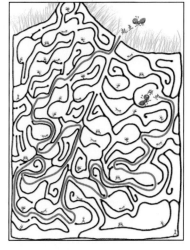

P43

59

P44

P45

P46

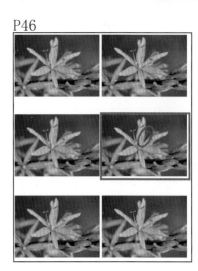

P49

1. 这样有斑点的狗有 9 只。（There are 9 of this kind of spotted dogs.）

这样有斑点的狗有 6 只。（There are 6 of this kind of spotted dogs.）

这样无斑点的狗有 7 只。（There are 7 of this kind of plain dogs.）

这样无斑点的狗有 8 只。（There are 8 of this kind of plain dogs.）

这样无斑点的狗有 5 只。（There are 5 of this kind of plain dogs.）

2. 有斑点的狗共有 15 只。无斑点的狗共有 20 只。所以，无斑点的狗比有斑点的狗多。（There are 15 spotted dogs, 20 plain dogs. The plain dogs are more than the spotted dogs.）

3. 这只狗与其他所有狗都不同。（This dog is different from the other dogs.）

P50

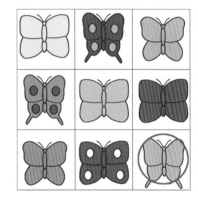

P51

属于春季 A 的有：3、8。
属于夏季 B 的有：4、7。

属于秋季 C 的有：1、5。
属于冬季 D 的有：2、6。

P52

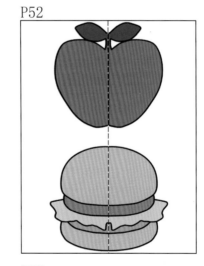

P53

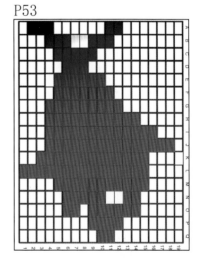

P54

P55

P56

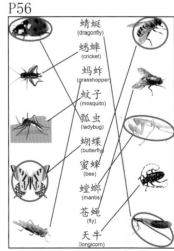